ANIMAL GROUPS

A TOWER OF GIRAFFES

BY SETH LYNCH

Gareth Stevens PUBLISHING

Please visit our website, www.garethstevens.com. For a free color catalog of all our high-quality books, call toll free 1-800-542-2595 or fax 1-877-542-2596.

Library of Congress Cataloging-in-Publication Data

Names: Lynch, Seth, author.
Title: A tower of giraffes / Seth Lynch.
Description: New York : Gareth Stevens Publishing, [2022] | Series: Animal groups | Includes index. | Audience: Ages 6 | Audience: Grades K-1
Identifiers: LCCN 2020028908 (print) | LCCN 2020028909 (ebook) | ISBN 9781538263631 (library binding) | ISBN 9781538263617 (paperback) | ISBN 9781538263624 (set) | ISBN 9781538263648 (ebook)
Subjects: LCSH: Giraffe–Juvenile literature. | Giraffe–Behavior–Juvenile literature.
Classification: LCC QL737.U56 L96 2022 (print) | LCC QL737.U56 (ebook) | DDC 599.638–dc23
LC record available at https://lccn.loc.gov/2020028908
LC ebook record available at https://lccn.loc.gov/2020028909

Published in 2022 by
Gareth Stevens Publishing
111 East 14th Street, Suite 349
New York, NY 10003

Designer: Tanya Dellaccio
Editor: Kristen Nelson

Photo credits: Cover Anup Shah/Stone/Getty Images; p. 5 Ketan Photography/Shutterstock.com; p. 7 tonyzhao120/Shutterstock.com; p. 9 Jeffrey B. Banke/Shutterstock.com; p. 11 Stock Pixels/Shutterstock.com; p. 13 Craig Fraser/Shutterstock.com; p. 15 Sergey Novikov/Shutterstock.com; p. 17 Lisa Hagan/Shutterstock.com; p. 19 Pat de la Harpe/Shutterstock.com; p. 21 XAVIER SANCHEZ CRIBALLES/Shutterstock.com.

Printed in the United States of America

CPSIA compliance information: Batch #CSGS22: For further information contact Gareth Stevens, New York, New York at 1-800-542-2595.

CONTENTS

Boldface words appear in the glossary.

Great Giraffes

Giraffes are easy to spot at the zoo! They're also the tallest land animals on Earth. They have a beautiful brown and white **pattern** on their fur. But these **mammals** are interesting because of more than their height and good looks!

There are four main species, or kinds, of giraffes. They all live on the **continent** of Africa. They live on **savannas**. There, they reach their long neck and tongues high to eat the leaves of the acacia tree.

Tall Towers

Giraffes are very **social** animals. They live in groups called towers. There are usually about 15 to 20 giraffes in a tower. Together, the group moves around a home range looking for food.

A tower is led by one adult male. Male giraffes are called bulls. The other giraffes in a tower are females, or cows, and their young. Female giraffes start to have babies, or calves, at about 4 years old.

Mothers and Calves

Soon after birth, calves are grouped together in a nursery. They play together and learn how to be part of a group. One mother stays behind to watch them. The other mothers leave them to look for food.

A calf stays with its mother for up to two years. Mother giraffes keep their young safe from predators. Female calves often stay with their mother longer. They also join groups near the home range where they grew up.

Male calves leave their mother to join another tower. This tower is made up of all young male giraffes. The young males take part in necking, or hitting each other with their heads and necks. This sets up a social order.

Giraffe Talk

Giraffes **communicate** with one another. They may roar, hiss, or grunt! Giraffes snort if they feel they are in danger. Mother giraffes might even whistle to call their calf. But some giraffe noises can't be heard by humans!

Giraffes in Danger

Giraffes are dying out. Human activities are causing giraffes to lose their **habitats**. Giraffes are hunted too. Their bones and fur are sold all over the world. Giraffes need to be protected so towers can continue to roam the savanna in the future!

GLOSSARY

communicate: to share thoughts or feelings by sound or movement

continent: one of the seven great landmasses on Earth

habitat: the natural place where an animal or plant lives

mammal: a warm-blooded animal that has a backbone and hair, breathes air, and feeds milk to its young

pattern: the way colors or shapes repeat

savanna: a grassland with scattered patches of trees

social: able and willing to form groups and live together

FOR MORE INFORMATION

BOOKS

Brinker, Spencer. *Baby Giraffes*. Minneapolis, MN: Bearport Publishing Company, 2021.

Emminizer, Theresa. *Giant Giraffes*. New York, NY: PowerKids Press, 2021.

WEBSITES

Giraffe

kids.nationalgeographic.com/animals/mammals/giraffe/

Find out more about the life of giraffes and see a video!

Giraffe

kids.sandiegozoo.org/animals/giraffe

Read tons of fun facts about giraffes here!

Publisher's note to educators and parents: Our editors have carefully reviewed these websites to ensure that they are suitable for students. Many websites change frequently, however, and we cannot guarantee that a site's future contents will continue to meet our high standards of quality and educational value. Be advised that students should be closely supervised whenever they access the internet.

INDEX